Bruno Guimarães Mothé
Fabrício B. Siqueira

Analysis of energy generation using an aeroderivative ethanol turbine

Bruno Guimarães Mothé
Fabrício B. Siqueira

Analysis of energy generation using an aeroderivative ethanol turbine

Feasibility study applied to the North of Rio de Janeiro

ScienciaScripts

Imprint

Any brand names and product names mentioned in this book are subject to trademark, brand or patent protection and are trademarks or registered trademarks of their respective holders. The use of brand names, product names, common names, trade names, product descriptions etc. even without a particular marking in this work is in no way to be construed to mean that such names may be regarded as unrestricted in respect of trademark and brand protection legislation and could thus be used by anyone.

Cover image: www.ingimage.com

This book is a translation from the original published under ISBN 978-613-9-63082-0.

Publisher:
Sciencia Scripts
is a trademark of
Dodo Books Indian Ocean Ltd. and OmniScriptum S.R.L publishing group

120 High Road, East Finchley, London, N2 9ED, United Kingdom
Str. Armeneasca 28/1, office 1, Chisinau MD-2012, Republic of Moldova, Europe
Printed at: see last page
ISBN: 978-620-7-76648-2

SUMMARY

Brazil is the future of humanity" José Walter Bautista Vidal

SUMMARY

STUDY OF THE FEASIBILITY OF GENERATING ELECTRICITY USING AN AERODERIVATIVE TURBINE POWERED BY ETHANOL IN NORTHERN FLUMINENSE

Brazil is one of the only countries that receives a lot of sunlight, as it is located in the tropics. Another characteristic of the country is that it has the largest share of fresh water in the world (12%), where this water is in a liquid state throughout the year, unlike countries located outside the tropics. In view of these particularities of our country and with a focus on the Norte Fluminense region, the purpose was to analyze the capacity of the Norte Fluminense region to plant sugar cane and produce ethanol for electricity generation or in the case of surpluses for export. The focus of this research is to highlight the importance of this region and introduce a pioneering technology in the world, which was implemented at Petrobras' Juiz de Fora Thermoelectric Plant, the result of a partnership with General Electric (GE), which developed a gas turbine that runs on flex fuel. By aligning this new management process, creating mechanisms for a more stable raw material in production, making ethanol the primary liquid fuel in Brazil and improving its transportation by means of ethanol pipelines, we can obtain a chain of interconnected suppliers, with the option of exporting surpluses, thus strengthening the ethanol industry and the fragile economy. We can see that there is no possibility of a project to create a network of ethanol pipelines and use them to power turbines for electricity generation unless there is a very high level of investment, which would be difficult if there were only public participation, i.e. in the situation of economic crisis in which the country finds itself, there would have to be private capital participation. It is a project of the highest value, since it expands the country's energy supply and develops an entire region. This new technology and form of production could take Brazil to the level of a self-sufficient country and energy exporter, which was the subject of this paper.

KEYWORDS: Ethanol. Biofuels. Alcohol product. Gas turbine. Energy.

1. INTRODUCTION

Brazil is the most suitable country for generating bioenergy. Its unique characteristics on the planet reveal its importance for the future, where there are resources that can prevent global warming, greenhouse gas emissions and also seek self-sustainable production, generating jobs, quality of life and avoiding the migration of people to large metropolises.

We need to guarantee the supply of ethanol so that the production system can be consolidated and remain in operation. In this way, we will evaluate the points of improvement that the installation of a pipeline system would bring, linking ethanol-producing plants directly to thermoelectric plants. By discussing the issues involved in the stages of bioenergy production and creating links between each point, we come to the conclusion that it is possible to produce more, in a more sustainable way, with greater operational flexibility, greater energy security, greater energy diversification and with the possible creation of a new market segment for ethanol in the country and abroad, making it possible to increase sales of the fuel and boosting the business of the sugar-alcohol sector in Brazil, provided that certain market parameters are reviewed and followed.

In view of the particularities of our country and the Norte Fluminense region, the focus of this research is to highlight the importance of this region and analyze the feasibility of introducing a pioneering technology in the world. This technology is the result of a partnership between Petrobras and General Electric, which developed a gas turbine that runs on dual fuel, natural gas and ethanol. This turbine was installed at the Juiz de Fora Thermoelectric Plant (UTE-JF).

2. OBJECTIVES

2.1 GENERAL

To study the feasibility of installing a thermoelectric power plant in the northern region of Rio de Janeiro, located in a seaport, to generate electricity using the primary fuel hydrous ethanol.

2.2 Specifics

Describe the history and growth prospects of the sugar-alcohol sector;

Describe how an ethanol-fired power plant works;

Estimating total bioelectricity generation capacity and simulating the profitability of selling surpluses in the north of Rio de Janeiro;

Explain the benefits of implementing a pipeline system linking the plant to the thermoelectric power station.

3. BACKGROUND

After the oil shock in the 1970s, the world realized that it needed to reduce its dependence on oil. This was followed by environmental conferences aimed at reducing greenhouse gas (GHG) emissions, and in 1997 the Kyoto protocol was launched. These were the two main events that preceded the search for a cheaper, renewable fuel with characteristics conducive to replacing fossil fuels (VIDAL, 2000).

The world is on the verge of two collapses: energy and the environment. It is becoming increasingly clear that the current model of energy production through fossil fuels is unsustainable. It is necessary to introduce new energy sources or improve current renewable energy sources. The environmental and social impacts caused by fossil fuels can be mitigated through the use of bioenergy.

The sun hitting the ground in Brazil is equivalent per day to the energy generated in 24 hours by 320,000 hydroelectric plants at Itaipu, the largest in the world (VASCONCELOS, 1998).

Brazil is the incomparable continent of the tropics, with more than 12% of the planet's fresh water, as shown in Figure 1, which is in liquid form throughout the year, unlike countries outside the tropics. The country also has a huge amount of arable land with high production potential. In addition, as it is located in the tropics, it has enormous sunshine, making it a country exclusively suited and predestined for the large-scale production of biomass, with a premeditated future for the great development of this energy (VIDAL, 2000).

Distribuição da água doce no mundo

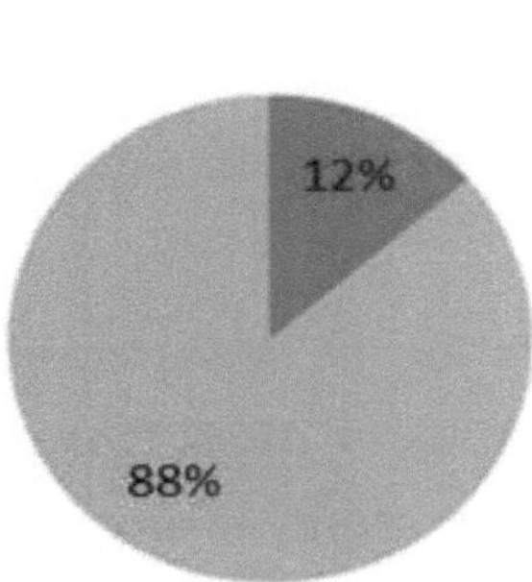

Figure 1. Distribution of fresh water in the world Source: SABESP

4. LITERATURE REVIEW

4.1 NATIONAL SUGAR-ENERGY INDUSTRY

The national sugar-alcohol sector has undergone profound changes in the last 30 years, which have affected all the economic segments related to it (NOVACANA, 2013).

The oil crisis was one of the reasons for the government to create the National Alcohol Program (PROÀLCOOL) in 1975 through Decree No. 76.593/75 - a new horizon in the Brazilian energy matrix, which caused the first major impact on the Brazilian sugar-alcohol sector, generating the first great wave of transformation (UNICA, 2013).

PROÀLCOOL aimed to create an alternative source of fuel for vehicles with Otto cycle engines and to promote job creation in Brazil. Despite its flaws, this program injected a considerable amount of capital into the sector, encouraging the start of technological development in all economic segments related to the production of alcohol from sugar cane (NOVACANA, 2013).

Brazil is the pioneer in the production of ethanol from sugar cane, both in the form of hydrous and anhydrous alcohol, and is competing with the United States for world leadership in the manufacture, export and consumption of this type of fuel.

Endowed with vast tracts of fertile land and a tropical climate suitable for growing sugar cane, Brazil enjoys a privileged position in the global biofuel market. Compared to the United States, Brazil's great advantage lies in its choice of sugarcane ethanol, which has an energy yield five times greater than corn ethanol, grown on the large farms of the American Midwest (FUSER, 2013).

A study by UNCTAD (2010) points to the need for investment in new units and adaptation of existing plants, as well as adequate regulation. A study by a UN body indicates that Brazil will have the capacity to produce 10 billion liters of second-generation ethanol by 2025, provided it invests in adapting and building new production units and has a regulatory environment adjusted to market circumstances. Unica, the association of sugar cane mills, took part in the study. The report provides a mapping of cellulosic ethanol projects and ventures in the United States, China, Canada and the European Union, divided by the current stage of operation, i.e. whether it is a pilot plant, a demonstration plant or in commercial operation, as well as the production capacity of each unit.

Brazil currently accounts for 12% of installed cellulosic ethanol capacity, corresponding to 177.3 million liters, compared to 34% in the United States (the world market leader) and 24% in China. Data from UNCTAD (United Nations Conference on Trade and Development) indicates that this mark will only be reached with the expansion of sugarcane crushing, modernization and integration of first and second generation ethanol production in existing plants, as well as the construction of 10 units per year exclusively for the production of 2G ethanol, starting in 2020 (BRASIL ENERGIA, 2016).

The technology currently used by the agricultural sector can be considered world-leading, especially that used in the central-southern region of the country.

According to data from the Ministry of Agriculture and Livestock, in March 2013, Brazil produced approximately 590 million tons of sugarcane, with the Center-South region accounting for approximately 90% of this production, as shown in Figure 2 (BIOSEV, 2013).

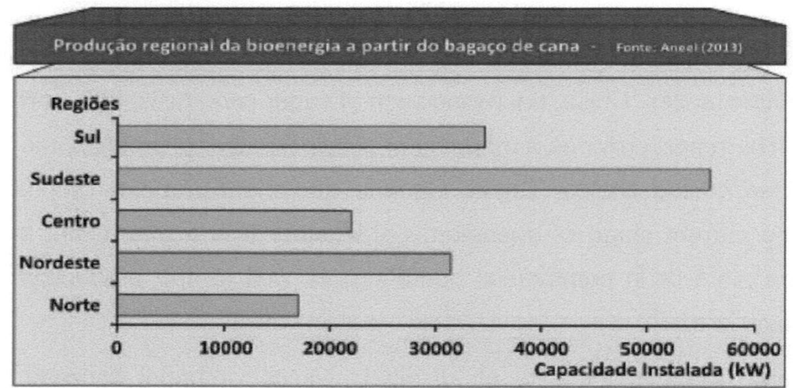

Figure 2: Regional production of bioenergy from sugarcane bagasse Source: Aneel 2013.

Today, the main source of energy from biomass residues in Brazil is sugar cane. As shown in Figure 3, 250 kg of each ton of the product is bagasse and another 204 kg is straw and tips. Everything can be reused to generate electricity.

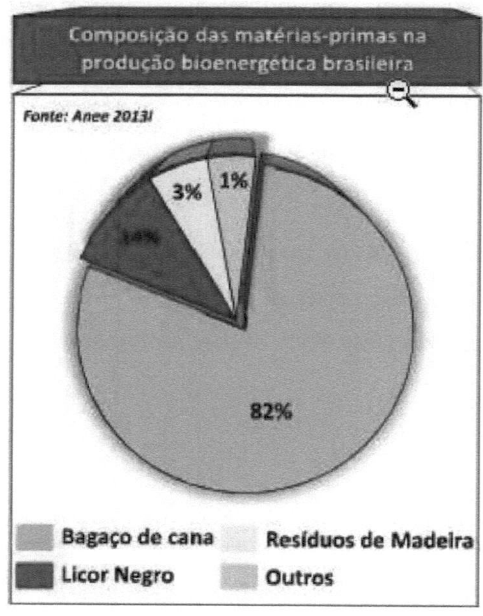

Figure 3: Composition of raw materials in Brazilian bioenergy production Source: Aneel 2013.

Sugar cane bagasse alone has the potential to generate more than 1.5

million kilowatts of electricity per year. The figures show greater potential if we take into account that of the 440 sugarcane mills in operation in Brazil, only 100 produce electricity for the national electricity system. According to the Acende Brasil institute, the sugarcane plantations in Brazil could generate around 14 million kilowatts per year (ÙNICA, 2014).

Recent studies carried out by the Center for Sugarcane Technology, Piracicaba / SP, show that there are several areas of Brazilian territory that could be used for the expansion of sugarcane plantations. The combination of three independent factors: the quality of the soil, the climatic conditions and the state-of-the-art technology developed in the agricultural field, have made Brazilian sugar cane one of the most promising sources of biomass, i.e. renewable energy, on the planet.

Auctions are the gateway for bioelectricity into the national electricity distribution network. In Brazil, the first new energy auction was held in December 2005.

According to CONAB data, the 2009/10 sugarcane harvest produced around 20,000 MWh (megawatt hours) of electricity, with just over 7,000 sold. This amount would be enough to supply approximately 28 million Brazilian homes, taking into account the total use of the energy sold by the mills. The annual production of bioelectricity from sugar cane already exceeds, for example, the electricity generated by Belo Monte, which would produce an average of 4,500 MWh of energy. There are estimates that, by 2021, bioelectricity will be able to produce energy equivalent to three Belo Monte plants (SOARES, 2015).

Despite being the tenth largest energy consumer, Brazil is one of the world's leading producers of hydroelectric power and biofuels, a renewable energy source.

Hydroelectric dams are still the most prominent in Brazil. However, a study published in 2014 by Oxford University shows that the construction of large hydroelectric dams is becoming economically unviable. The reason for this is the skyrocketing budgets and delays in construction.

An initiative to publicize bioelectricity, launched in 2011 by the AGORA Project, identified the immense growth potential of this form of energy generation. By 2021, bioelectricity supplied to the national grid alone could account for up to 18% of the Brazilian electricity matrix.

Figure 4 shows the projections concluded in 2012 by the Union of the Sugarcane Industry (ÙNICA), which define the pace of growth that will need to be observed in the Brazilian sugar-energy sector.

Figura 4. Challenges for the growth of the sugar-energy sector until 2020 Source: Unica.

Table 1 shows some parameters or indicators of the performance of the industrial area of the sugar and alcohol production sector at different times: The start of PROALCOOL, when the sector's major technological development effort began, and the present day.

Table 1 - Industrial Performance Indicators

Description	Unit	Home Proàlcool	Today
Cane crushing capacity in 6 2000mm (78")	Ton/day	5500	14000

mills Extraction of juice in 6 mills as a percentage of cane sugar	%	93	97
Average fermentation time	Hours	18 a 24	4 a 6
Amount of sugar in the juice transformed into alcohol during fermentation	%	80	91
Loss of alcohol in the distillation process	%	2	0,5
Steam consumption in sugar production per ton of cane processed	kg steam / t. cane	600	320 (1)
Steam consumption in the distillation of anhydrous alcohol	kg steam / liter alcohol	4,5	1,8 (1)
Potential methane obtained from vinasse	Nm^3 gas / liter alcohol	0	1
Maximum alcohol production per ton of cane - (13% cane pulp)	l. alcohol / t. cane	66	86
Maximum bagasse surplus in an alcohol plant in relation to the bagasse generated	%	Up to 8	Up to 78
Steam generation pressure in boilers	bar	22	92 (1)
Electricity production capacity per ton of sugarcane crushed burning only bagasse	kWh	2,7	80
Cost of producing anhydrous alcohol at the mill (2)	USD / m3	700	200

Source: (I)Dedini / (2)Datagro Conf. Int. 2001/ International Sugar Journal - ISJ

The figures in Table 1 are strong indicators of the technological development that has taken place over the last 30 years in the sugar-alcohol sector, as well as its competitiveness at world level. The factors driving this development are: initially, the expansion of domestic consumption of fuel alcohol, followed by the increase in international demand for sugar and, currently, the beginning of the full use of sugar cane as a renewable energy source (SOARES, 2015).

The sugar-alcohol sector is going through an economic crisis of great intensity and certainly the most persistent and long-lasting since the end of the liberalization process in this sector at the end of the 90s.

As an agro-industrial activity, the sugar-alcohol sector is directly influenced by an intrinsic characteristic: the seasonality of raw material production. This peculiarity affects most plantations and imposes a strong constraint on the marketing of products derived from these crops. Other factors such as the international economic crisis, which came to a head in August 2008, resulting in higher interest rates and less availability of bank credit lines, which increased the liquidity difficulties of many economic groups; the management problems of some groups which took on debts disproportionate to their capacity in an attempt to speed up production growth and increase their market share; the pressure of supply caused by units needing to restore their cash flow and liquidating their production in a disorderly fashion at whatever prices are possible (CONAB, 2010).

The cost of producing bioelectricity is still very high, which makes it difficult to consolidate as an energy source. One possible way out would be greater government incentives in relation to the tax burden, investments in infrastructure and a new calculation of the price purchased by the government (NOVACANA, 2015).

Biomass is essential for the future of Brazil's energy matrix. Its participation means less dependence on fossil fuels and security of supply. Brazil has the technological capacity, experience and raw materials with the potential to supply. However, incentive policies must be applied, given the delicate moment that Brazilian ethanol is facing.

4.2 Historical dependence on oil

Modern society has become increasingly dependent on energy in almost all its activities. This is an ongoing historical trend.

From the end of the 18th century, with the industrial revolution, mineral coal became the main source of energy, until it was replaced in the 20th century by oil, which has driven the gigantic economic and social transformations associated with the modern era more than any other natural resource. Oil was the main item in the global energy matrix throughout the 20th century and, according to all reliable forecasts, will retain this position for a long time to come, despite all the efforts to develop other fuels.

Historically, the introduction of fossil fuels coincides with the emergence of modern industries, organized according to the capitalist economic model.

Why has energy acquired such a decisive weight on the global and regional political stage? The answer can only be found in a historical perspective. All the economic progress of the 20th century, and also that which is projected for much of the 21st century, took place on the basis of a clearly defined material foundation - the wide availability of energy resources, especially oil and natural gas, obtained in large volumes and sold, for the most part, at very low prices. Similarly, energy was also of critical importance in the military sphere. Historians are unanimous in including, among the decisive factors for the Allies' victory in the Second World War (1939 - 1945), their access to almost unlimited stocks of oil, a strategic fuel present in large reserves in both the United States and the Soviet Union. Their enemies, Germany and Japan, had no such advantage.

The particular geographical nature of hydrocarbons (oil and natural gas), whose reserves are distributed very unevenly across the planet, underlines their importance in the sphere of international relations. States generally see guaranteeing the supply of these fuels, especially oil, as a priority security issue, subordinate only in the scale of concerns to national defense.

The "war for resources" theorists are convinced that market forces alone are incapable of resolving the imbalance between supply and demand, which can lead some states to pursue their goals through force or the threat of force.

The advance of industrialization in the 20th century triggered a new race for control of sources of raw materials. Among these sources was oil, which proved decisive in the outcome of the two world wars.

All projections indicate that oil will maintain its dominant position as the main commercial source of energy until 2030, although gradually reducing its share of the total supply.

4.3 History of the Sugarcane Industry in Northern Rio de Janeiro

In 1538, Pero de Góis founded a small settlement which he named Vila da Rainha. It was located a short distance from the mouth of the Itabapoana River, formerly known as the Managé River, in the present-day municipality of São Francisco de Itabapoana, one of the oldest in São João da Barra. In the North of Rio de Janeiro (NF), the donatary planted the first sugar cane seedlings and built water-powered mills, with the intention of producing 2000 arrobas of sugar a year (OSCAR, 1985, p. 39-40 apud PESSANHA, 2004, p. 31).

After Pero Góis' unsuccessful attempt to set up sugar mills in São João da Barra, the sugar industry would resurface in the town of São Salvador (Campos dos Goytacazes) in the second half of the 17th century, when its dynamism would be ensured by strong demand from the international market. However, it wasn't until the 18th century that sugar cane cultivation took over from livestock farming on the Campos plain.

From the beginning of the 17th century, the first settlements appeared in the NF region, which would later define cities such as São João da Barra, Campos dos Goytacazes and Macaé (ARRUDA, 1986, p. 87 and 95 apud PESSANHA, 2004, p. 33).

Table 2 shows the importance of the economy of the north of Rio de Janeiro, which can be measured by the significant number of sugar mills it had at the time and the consequent production of sugar, and by the considerable number of slaves needed for production.

The formation of the agro-sugar economy in the NF is characterized at the end of the 18th century and the beginning of the 20th century by two key reasons which distinguished the production process in this region from that in the Northeast of the country, as shown in Table 3.

Table 2 - Number of mills in the Southeast - Late 18th century

Southeast region	Sugar mills	Brandy mills
Guanabara	228	85
Angra dos Reis and Ilha Grande	39	155
Cabo Frio	25	9
Campos dos Goytacazes	324	4
São Paulo	400	550
Total	1.016	803

Source: (NETO; PESSANHA, 2004, p. 41).

Table 3 - Number of mills in the Northeast, end of the 18th century

Region	Sugar mills
	37
Paraiba Pernambuco	296
Alagoas	140
Sergipe	260
Bahia	260
Total	993

Source: (NETO; PESSANHA, 2004, p. 41).

As table 4 shows, from 1820 onwards, the Campinas economy would undergo changes. Only those mills that had large reserves and huge financial resources to buy and install machinery, to adopt new production systems and to buy more slaves from traffickers based in São Joâo da Barra or Macaé would survive (OSCAR, 1985, P. 105 apud PESSANHA, 2004, p. 42).

Table 4 - Number of engenhocas, steam mills and mills in Campos dos Goytacazes (19th century)

Years	Gadgets	Steam engines	Power

17

			plants
1827	700	1	
1852	307	56	
1861	267	68	
1872	207	113	
1881	120	252	5

Source: (NETO; PESSANHA, 2004, p. 42).

Thus, the mills underwent a process of mechanization that accelerated the division of labour, allowing for increased productivity and the beginning of the industrialization process in the production of sugar and sugar derivatives in the region. On September 12, 1877, the first central sugar mill in Latin America was inaugurated in what is now the municipality of Quissamã. In the following years, the following mills were inaugurated: Central de Barcelos (1878), Queimado (1879), Capim (1881) and Usina do Limão (1879), in Campos (ANDRADE, 1994, p. 71 apud PESSANHA, 2004, p. 42).

The 19th century was unique for the sugar industry in Campos dos Goytacazes. It saw the disappearance of the engenhocas and traction mills. We saw the emergence of steam mills and their replacement by central mills (central mills had the sole task of milling sugar cane and could not produce sugar cane). The activity of production fell to the farmers) (PINTO, 1987 apud PESSANHA, 2004, p. 48).

Table 5 - Sugar production in Campos dos Goytacazes (1830 - 1929)

Period	Tons (t)	Thousands of 60kg bags
1830	10.000 t	166.666
1835	12.240 t	204.000
*1852-1861	8.798 t	146.633
*1862-1871	13.684 t	228.065
*1872-1881	16.757 t	279.277

1882	21.350 t	355.830
1893	16.102 t	268.363
**1900-1905	21.753t	362.558
**1906-1910	26.203 t	436.724

*Gathered production in ten-year averages **Gathered production in five-year averages

Source: (NETO; PESSANHA, 2004, p. 48).

The astonishing growth in sugar production at the end of the 19th century and the beginning of the 20th century pointed to radical transformations in the region's sugar production process. With the emergence of the mills, highly mechanized industrial units with large-scale production, the average productivity of the North of Rio de Janeiro increased significantly, making it very competitive with the other producing regions of the country, mainly located in the state of São Paulo and the Northeast (SOUZA, 1935, P. 77 apud PESSANHA, 2004, p. 49).

However, with the crisis of 1929 and the revolution of 1930, important changes took place in Rio's economy (CME, 1933, apud PESSANHA, 2004, p. 59).

At the beginning of the 20th century, Campos dos Goytacazes had 27 mills and sugar production was growing. The five largest mills in terms of production were, in order, Cupim, Mineiros, Santa Cruz, Tocos and Barcelos (PINTO, 1987, p. 78 apud PESSANHA, 2004, p. 62).

Table 6 - Sugar production in the 20th century - North of Rio de Janeiro

Year	Thousands of tons	Year	Thousands of tons
1917	19,5	1959	102,6
1920	35,0	1961	124,2
1930	22,4	1963	90,3
1932	24,7	1965	132,9

1934	30,4	1967	136,6
1936	43,5	1969	128,8
1937	41,8	1971	123,1
1939	38,4	1973	169,6
1941	52,7	1975	150,2
1943	40,8	1978	157,8
1945	58,2	1980	120,4
1948	65,6	1983	121,8
1950	64,1	1986	123,9
1954	82,2		

Source: (NETO; PESSANHA, 2004, p. 63).

The global economic crisis of 1929 led to a downturn in world sugar consumption, affecting the flow of production from the north of Rio de Janeiro. With the 1930 revolution, the Sugar and Alcohol Institute was created, whose main role was to control sugar and alcohol production by imposing production quotas.

The IAA policy had introduced the blending of alcohol into gasoline and supported a growing potential market.

From the end of the 1950s onwards, many mills in Campinas began to be acquired by mill owners from the Northeast. São Paulo had gained a foothold in the national and international sugar and ethanol market, which helped make it the country's largest sugar producer. Campos dos Goytacazes then began to suffer the competitive influence of other states and lost important market shares in the domestic sector of the Brazilian economy (PINTO, 1987, apud PESSANHA, 2004, p. 63).

The economic and social crisis that hit the country at the beginning of the 1960s, coupled with the depressed price of sugar, had a negative impact on the economy of the north of Rio de Janeiro, which deepened after the military coup of 1964.

The peak in the effective production of sugar and alcohol came in the second half of the 1970s, especially due to the government's stimulus to the sugar and alcohol industry through Proàlcool.

However, the 1980s were a critical decade for the sugar economy in the NF. The low growth rates of the national GDP, in the midst of the country's inflationary spiral, made private credit for new investments unfeasible and made producers more dependent on the increasingly precarious subsidy policies of the federal government. The negative impact on the region is therefore understandable, as the sector decreased in production at the end of the 1970s and collapsed in the 1980s and 1990s. The most visible results of the sector's economic stagnation and its impact on the region were the closure of numerous mills, the high level of unemployment generated and the precariousness of labor relations (SILVA, CARVALHO, 2004, p. 65).

4.4 Food safety associated with biofuel production

The production of biofuels, such as ethanol and biodiesel, has been criticized and mistrusted by various authors, and is always questioned as causing a reduction in land or food production.

Just as technology is advancing in industry, it is no different in the countryside, where universities and research centers have been improving the way they plant, harvest, fertilize and so on, in order to increase production, both for food and biofuels.

Insured in accordance with the Universal Declaration of Human Rights (UN, 1948), some companies have placed this precept as the basis for their guidelines or as part of a social responsibility program so that biofuel production does not affect food production.

In an interview with the magazine "Especial Pâginas Verdes N° 04 June 2014", Silvano Cavalcante, Production Coordinator Semiàrido Norte II at Petrobras Biocombustiveis said:

When Petrobras decided to invest in biofuels, there was a lot of controversy and one of the arguments against it was the fear that the production of raw materials for biofuels would reduce food production. In practice, the opposite has happened: food production has increased, including corn, which has also increased livestock farming. This has been intensifying since 2004, with the creation of the National Program for the Production and Use of Biodiesel (PNPB) and, later, with the promulgation of the Ministry of Agrarian Development's Ordinance that created the Social Fuel Seal, regulating that biodiesel producing companies allocate, on average, up to 30% of their investment to include family farming in the production chain. (BIOCOMBUSTIVEL; SMES, 2014, p.3-4)

Also in the interview, he mentions that Brazil has a surplus in corn and bean production, crops that are encouraged to be planted among castor bean crops, which are exported. In addition to this, there is also the issue that the diversification of crops in small plantations (sub-existence) leads producers to regularize their profits and, in the event of prolonged droughts, not be left without a source of income, since oilseed crops are more resistant to droughts and pests. In short, planting crops for biofuels not only strengthens the economy and small farmers, but also prevents migration between cities and regions in the country. It is important to note that the total use of arable land for energy purposes is minimal. In fact, only around 1% of the world's arable land is currently used for the production of liquid biofuels, with prospects of this rising to 3% to 4% by 2030 (BFS/FAO, 2008, *apud* BNDES, 2008, p. 253). Similarly, it's hard to believe that there are effective surface area restrictions for producing food and biofuels when you consider that the areas currently under cultivation across the planet (around 1.5 billion hectares) represent approximately 12% of arable land. Furthermore, a significant proportion of current grain production is destined for animal feed, meeting the food needs of the world's population in a very asymmetrical way (BNDES, 2008). In conclusion, as the area currently occupied by crops for energy production is small, the negative effect on world food supply and/or access is minimal, if not nil. The problem in some countries where deaths from hunger occur may be associated with other factors, such as low land productivity or a lack of subsidies for small producers, where the quickest and most widely used solution in the most fertile fields is regular land maintenance and investment in productivity.

4.5 Available sugarcane planting area

For a territory to be considered suitable for planting, it must have a minimum of water, sunlight and carbon dioxide, while for bioenergy production it must also have fertile soil and topography and, in addition to these requirements, some mineral nutrients such as nitrogen, phosphorus and potassium and, to a lesser extent, boron, manganese and sulphur.

Brazilian land is very favorable for planting sugar cane and oilseeds such as castor beans and soybeans, as it is rich in these nutrients by nature. In addition to this initial point, we already have mastered technology for replenishing these nutrients through chemical fertilizers, and for sugar cane, the washing residue itself (Vinhaça or vinhoto) and the ash from the boilers, when returned to the plantation land, has a dual function, the replenishment of nutrients (in greater quantity potassium and phosphorus and to a lesser extent other nutrients), and that of a natural insecticide, avoiding aggression towards fauna and flora. In average Brazilian conditions, the application of vinasse and filter cake, although it does not have a significant effect on the supply of nitrogen, reduces the demand for phosphorus (P_2O_5) from 220 kg/ha to 50 kg/ha and potassium (P_2O) from 170 kg/ha to 80 kg/ha, maintaining similar levels of productivity (CGEE, 2005, *apud* BNDES, 2008, p.189).

With regard to topography, much of the country's land has a low slope and, in particular, the NF region, which will be the subject of this study, is located in the coastal region and offers exclusive planting conditions, facilitating mechanized harvesting and avoiding land erosion.

A study carried out by Hoogwijk *et al.* (2003), delimited the cultivable areas on the planet, where 13.2 billion hectares are divided into 1.5 billion hectares that are currently used for the production of food for humans and animals, corresponding to 11% of the total.

Figure 5 shows the distribution of land use across all continents, signaling the existence of areas available for the expansion of agricultural frontiers and the production of bioenergy, especially in places that are still little explored or used extensively, such as low productivity pastures.

Figura 5. Distribution of arable land on the planet Source: (BNDES, 2008).

Nevertheless, most of Brazil's pastures have low productivity, and in particular, areas that are resting or not being used, as shown in figure 6 below:

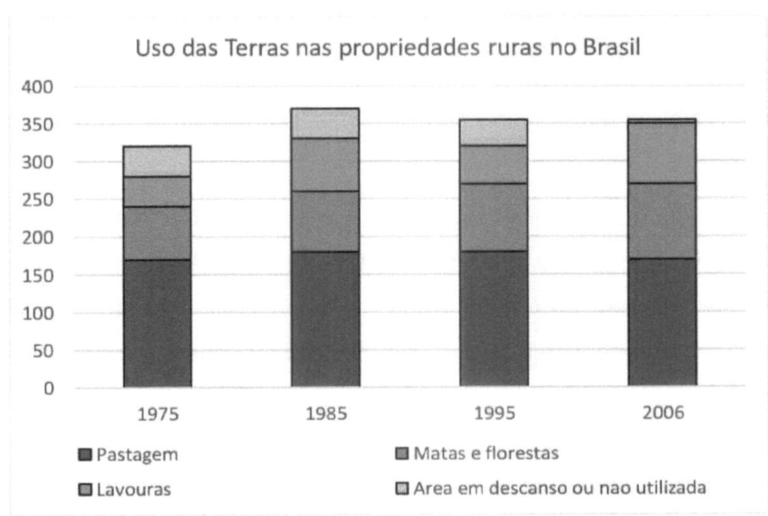

Figura 6. Land use in rural properties in Brazil Source: (BNDES, 2008).

Due to its continental dimensions, Brazil has a great diversity of soil classes and climatic types that result in a great variation in the productive potential of its land. This diversity, combined with other factors such as slopes and protected areas such as the Amazon, Pantanal and other reserves, creates a need to map and control the areas where planting is permitted.

Vidal (1995) carried out surveys and estimated that by growing sugar cane or cassava on 1% of Brazil's territory, 800,000 barrels/day of ethyl alcohol could be produced. Putting the figures in a table and comparing them with the ANP's 2014 data, we have:

Table 7 - Total land area for ethanol supply

	Thousand M^3
Production in 2014	28.820,34
Production in 1% of the territory	46.424,29

Source: Own authorship.

The 2014 production figures are for anhydrous and hydrous ethanol, while the 1995 figures are for hydrous only. Another variable to consider is that more than two decades have passed since that survey and since then, many improvements have been made in the field, increasing productivity even more for this territory of 1%.

Although Brazil has a strong agricultural sector, we can see that it is possible to increase the amount of land used to generate bioenergy without damaging food, pastures, forests and protected areas, only by making better use of unused land or, if combined with research into productivity improvements and genetic improvements, Brazil has the potential to become the largest exporter of energy in the form of biomass.

4.6 Characteristics of the sugar cane input

In order to discuss the feasibility of setting up the complex industrial park we are studying, it is of fundamental importance to analyze the input, which is sugar cane. The definition of the plant, as described in the National Energy Plan 2030:

It is a herbaceous, cespitose and perennial plant, from the grass family, which needs deep and fertile soil, a well-distributed minimum rainfall of between 1,200 and 1,300 mm/year and temperatures of between 20 and 24°C, and does not tolerate frost. It should only be harvested during certain periods of the year because of its maximum sugar content. In general, after the first harvest, sugar cane will grow again, producing another 4 or 5 harvests a year with decreasing productivity. On average, during an entire production cycle, productivity varies from 50 to 100 t/ha.year depending on the agricultural practices adopted. In Brazil, sugar cane can be produced in almost all parts of the country, especially between parallels 8° and 24°. (ENERGETICA; ENERGIA, 2007, p.49)

For sugar cane to acquire the ideal amount of sucrose, it is necessary for the land to have at least two ideal seasons, one dry with cold characteristics and the other hot and humid, thus helping germination, tillering and vegetative development. The ideal planting season for the Center-South region is from January to March, while in the North-Northeast it is from May to July. These planting times, in the current situation of harvesting and processing, if integrated into the energy matrix, complement the country's energy generation in a unique way, because at the time when we have the lowest hydroelectric generation, we would have generation through bioenergy supplying the needs.

Table 8 - Technical information on sugar cane cultivation

Item	Data
Cycle	5 years
Number of cuts	5 cuts
Sugarcane productivity	85 ton/ha (120 65)
Sugar yield	138kg/ton
Alcohol yield	82L/ton

Source: (MINISTÉRIO DA AGRICULTURA, PECUARIA E ABASTECIMENTO, 2007, p.11).

BEN data shows that in 2013 renewable energies accounted for 41% of the national domestic supply, and sugarcane biomass represented 16.1% of this total, surpassing all other forms of renewable energy.

4.7 Advantages of using ethanol compared to gasoline

This section will discuss the advantages, improvements and other benefits of using alcohol to replace fossil fuels. The hydrocarbon residues from burning in internal combustion engines for transportation, energy generation, etc. compared to the residues from the same processes but using ethanol are more aggressive both to the environment in general and to human health. Below are some points highlighting the difference between hydrocarbons and ethanol:

4.7.1   Transportation and Storage

Transportation and Storage: Oil exploration throughout its history has resulted in a large number of accidents, most of them of major proportions. When oil or its derivatives contaminate waters, whether surface or deep, or enter the human food chain, they can have adverse effects on health.

Ethanol, due to its chemical structure and greater capacity for degradation in the environment, represents zero risk of the alterations described above. The greatest risk related to ethanol is inhalation or sudden contact. Therefore, having correct supervision in order to avoid more serious accidents would be enough.

Because they are liquid fuels, gasoline and ethanol basically don't differ in their transportation and storage systems, but three aspects are more influential in ethanol, as highlighted below:

However, there are at least three particular and important factors to consider: the seasonality of ethanol production, the spatial dispersion of this production and the compatibility of the materials of the tanks and pipes that will be in contact with the ethanol and its mixtures. (BNDES, 2008, p.60)

These aspects will minimize the negative effects of an alcohol pipeline, such as the one being built in São Paulo by LOGUM, where an extensive network of pipes flows production to places where it is stored in tanks.

By taking advantage of this pipeline system to interconnect the alcohol-producing plants and transport production more reliably and with less downtime and setup time, the alcohol pipeline has the potential to supply the thermoelectric plant and also make it easier to export the alcohol produced in addition to domestic consumption.

According to Ballou (2006), the movement of products via pipelines is very slow, at no more than three to four miles per hour. On the other hand, it is 24 hours a day, seven days a week, which makes the effective speed much higher when compared to other modes.

The pipeline pumping system is the most reliable, with the lowest operating costs and the lowest risk of loss of production or accidents.

He also adds that the damage and loss of products in pipelines is low, as can be seen in Table 10. Liquids and gases are not subject to damage to the same degree as manufactured goods. Furthermore, the number of risks that can affect a pipeline operation is limited.

Table 9 - Relative ranking of transport modes by cost and operational performance characteristics

Performance characteristics

		Delivery Time Variability			
Mode of Transportation	Cost	Average time Delivery faster	Absolute 1=	Percentage	Losses and Damage
Railroad	1 = higher	3	1 = smaller	1 = less	1 = smaller
	3		4	3	5
Road	2	2	3	2	4
Watercraft	5	5	5	4	2
Pipeline	4	4	2	1	1
Air	1	1	1	5	3

Source: (BALLOU, 2006, p.58).

4.7.2   Health

The burning of fossil fuels in internal combustion engines results in greenhouse gases (GHG). In gasoline engines, there is a family of volatile compounds and polycyclic aromatic hydrocarbons, which have great toxic and carcinogenic potential. Activities related to oil exploration and refining are harmful to human health. According to the book Ethanol and Bioelectricity, several epidemiological studies have reported an increase in cases of respiratory and cardiovascular diseases and tumors (leukemia and neoplasms of the central nervous system) in the vicinity of refining and petrochemical areas.

In ethanol engine emissions, the organic compounds are 70% ethanol and 10% aldehydes.

Aldehydes are highly reactive organic substances containing a carbonyl group (double bond between carbon and oxygen atoms), which have a high affinity for lipids, proteins and DNA (COMEAP, 2000, *apud* MACEDO, I. C.;

SOUZA, E. L. L., 2010, p. 111).

In the case of ethanol, harvesting must be mechanized to avoid the adverse impacts of emissions from burning (Ribeiro H., 2009), given the body of evidence on the adverse impacts on the health of workers and the surrounding population.

The equation below shows the chemical reaction that takes place in the ethanol combustion process, which is composed of $C\ H_{25}\ OH$. Thus we have:

$$C_2H_5OH + 3O_2 \rightarrow 2CO_2 + 3H_2O \tag{1}$$

The chemical reaction of ethanol combustion, due to its composition, favors the formation of water, since the reaction of oxygen with the internal hydrogen of the fuel generates bound water. Under these conditions, the burning of ethanol combined with atmospheric air, which contains 21% oxygen and 78% nitrogen, forms nitrogen oxide. The formation of nitrogen oxide in the ethanol combustion process, due to the presence of water, occurs in smaller quantities when compared to the burning of natural gas.

Therefore, the smaller the amount of nitrogen oxide released into the atmosphere, the smaller the possibility of polluting gases being formed (MACHADO, 2010).

4.7.3 Global Warming

The issue of heating also influences the topic of health above. In addition to the items that will be mentioned, it is also worth remembering that it further compromises the health of the population in general, increasing the number of hospitalizations, deaths and others.

Ethanol has several advantages over fossil fuels in terms of GHG emissions.

A biofuel such as ethanol, because it is more neutral (from the moment the plant captures carbon dioxide from the atmospheric air, the carbon dioxide in the exhaust after burning is better balanced), in terms of GHG emissions, when compared to fossil fuels, can help to reduce impacts. The main aspects are food security, scarcity of water resources and thermal stress.

Food security is the most obvious issue, given that if the planet continues to warm up, some arable areas could suffer from desertification (in the case of Brazil, the most likely is the semi-arid northeast), thus causing a greater problem of social and economic inequality. As for water resources, in coastal regions, the rise in sea level means that aquifers are likely to become salinized, reducing the quantity and quality of water. Climate change is causing a shift in the rainfall cycle, where at times of the year it rains too much, causing flooding, and at other times prolonged drought. Excessive rainfall, causing flooding, compromises cisterns and dams, thus jeopardizing supply, and bringing contamination to reservoirs by human and animal waste, causing a chain of more diseases.

All these situations, in which global warming does not cease, are expected to increase morbidity and mortality from water-borne diseases and force the population to migrate to the affected regions.

4.7.4 Work and Income Division

Division of labor and income: A characteristic of the oil industry is the concentration of income and the poor distribution of labor, given that, compared to other industries, it requires a smaller contingent, with higher salaries due to the need for skilled labor. The sugar cane processing industry has other characteristics which are a better distribution of income and work. Generally, the workers who work in the fields (cutting) have the lowest salaries and the lowest level of education, which correlates to lower salaries. If there were incentives for mechanized cutting and harvesting, these jobs could be reallocated to other stages of the process. It also helps to reduce the concentration of populations in the major centers, reducing migration to the metropolises, where due to the large concentration of people, life loses the quality of the countryside, burdening the public budget with more costs and expenses.

The Brazilian experience shows that the activity generates 835,000 jobs, where salaries are among the highest in the agricultural sector (second only to soybeans).

4.8 Consolidation of ethanol as a primary liquid fuel

If we look at the history of how the international commodities market has developed, we can see that everything starts from the principle of demand and consumption, but there are many obstacles on the road to the globalization of a commodity, where in order to be consolidated as a commodity, the product must be a global necessity, or at least be of interest to nations with great influence on a world scale. Another point is that it must be a product that can be easily standardized, or that can be segregated by type, so that it can be sold at different prices in the case of different production processes and added values. Once these requirements are satisfactory for the product, the focus will be on drawing up norms and standards for production and product standardization.

The most common commodities are agricultural and energy, where oil has a unique potential, with its own stock exchanges and markets, and is considered a political commodity. In the face of the world's growing need for energy, civilization has sought out the most convenient forms of energy in the short term, at the whim of large corporations, where oil has fitted in perfectly. As it is a product that can be found in the vast majority of countries, but in varying quantities (production feasibility), the world has prepared itself to use it on a large scale. As it is a fossil product (finite), there are two major collapses in production in sight: availability to meet world demand and the environmental crisis.

Throughout history, oil has been used as a means of dominating territories and political influence, where the countries that dominate production, not necessarily the producers, hold this form of energy as an instrument of hegemony. As soon as this dominance ceases, or a strong intervention takes place, as happened in the 1970s, when the major producers decided to limit production, leaving the world in deficit, projects for alternative forms of obtaining energy become feasible, as was the case with PROALCOOL in Brazil. A search for substitutes for oil began, with heavy investment in the research and development of renewable fuels, making the Brazilian program the most complete biofuel program in the world. However, a program of this magnitude was only possible because there was government intervention, with incentive policies and price controls added to the involvement of the private sector, creating the conditions for the emergence of an industry capable of competing with gasoline (KLOSS, 2012).

There are various ways of promoting the consumption of biofuels, such as instruments that reduce the price in order to make them more economically competitive, which can take the form of direct tax exemptions on fuels, or indirectly on production inputs such as energy, water, fertilizers, or even tax credits for the purchase of physical structures (KLOSS, 2012).

Notably, a "new" industry finds it very difficult to enter the world market, and in the ethanol scenario, where production is most efficient in the tropics (a region predominantly made up of underdeveloped or emerging countries), it becomes more complicated for these forms of energy to become commodities.

Recent actions have been taken to try to circumvent the barriers imposed, but since the largest ethanol producers process them from different raw materials, the standardization of the product, the price and the influence of non-market issues become "internal" obstacles.

Another fact that can be considered for ethanol to consolidate is the integration of the production chain by controlling supply and distribution activities, with the aim of increasing efficiency and isolating price volatility in markets. By comparison, the oil industry had an example of integration in this way, and it was such an effective process that it made Standart Oil Co. (founded in 1865 and in 1879 already accounting for 90% of refining capacity) a monopoly, but it was characterized as illegal and was dissolved into several companies. In Brazil, Petrobras was created by the state with the justification of increasing the supply of the product and lowering its cost, where it was the only company authorized to prospect, research and refine oil until 1995, after which it was opened up to private foreign capital (KLOSS, 2012).

Ethanol's entry into the international market has been in the form of ethanol blended with gasoline, in some cases as a substitute for Methyl Tertiary Butyl Ether (MTBE) because it is a volatile compound that causes serious problems for human health, or as a government project to introduce alcohol into the energy matrix to make it less aggressive to the environment, as well as being less damaging to internal combustion engines compared to its recessor.

Several countries have been progressively adding ethanol to their liquid fuels, but not all countries are able to meet their current or future needs, so there is a possible gap for ethanol exports. Some countries, in order to protect their local industry, add taxes to ethanol imports, but even so it is possible to see how promising Brazilian ethanol would be if it were integrated into the international market, as Kloss mentions:

Brazilian ethanol, even with the addition of import duties, freight and insurance, reached refineries on the East Coast (USA) at a price of US$1.75 per gallon, while the price of corn ethanol produced in the Midwest averaged no less than US$1.90 per gallon, i.e. 8.5% higher than the Brazilian competitor. (KLOSS, 2012, p.103)

Brazilian ethanol is overcoming the barriers imposed by the world's largest producer (USA), even after heavy taxation to protect the local industry. There is no greater barrier to native ethanol consolidating itself as a renewable, inexhaustible raw energy material and a solution to the world energy crisis, and these come in the form of fiscal, tax and marketing barriers, because there is no technological limitation to its widespread use in the transport sector, and now with cogeneration on the rise and conversion via an aero-derivative turbine, it has unparalleled potential.

In 2009, the EU set a global target of a 20% share of renewable energies in its energy mix as part of the plan to mitigate the effects of climate change, with the transport sector contributing 10% through the use of biofuels and other renewable fuels.

The European Community, even though it is the third largest producer, explicitly recognizes that it is unable to supply the quantities of biofuels - ethanol and biodiesel - required by the European mandate through domestic production alone; under these conditions, imports from third parties will continue to be necessary. The EU authorities are working with a scenario of 50% biofuels produced in EU countries and 50% imported (KLOSS, 2012, p.72).

In the case of Brazil, the Mercantile and Futures Exchange (BM&F) has favored the expansion of agricultural production. The Brazilian exchange is one of the few in the world to have a futures contract for ethanol, which is favorable to developing countries, as commodity producers do not normally have the economic conditions to trade on foreign exchanges. The contract in this form of exchange standardizes parameters for anhydrous ethanol, such as appearance, color, total acidity, electrical conductivity, specific mass at 20°C, alcohol content, ethanol content, hydrocarbon content, amount of copper and the corresponding methods of analysis (KLOSS, 2012).

However, all these changes, plans and strategies won't be enough if there isn't an effort on the part of the countries that are leaders in production or are interested in renewing their energy matrix. There should be a mutual effort to formalize ethanol as an energy commodity at a global level. As a premise for this, standards and standardization institutes for ethanol should be created. The barriers exist, and they lie in the need to standardize the product, since each country has a plant that is best adapted to its region. These crop variations bring diversification in the quality of the end product, not only for ethanol, but also for biodiesel.

There are several reasons to believe that ethanol could replace gasoline and fossil fuels on a global scale, both on the supply and demand side. Few developing countries are self-sufficient in oil, and dependence on this energy source results in the transfer of resources that could be converted into urgent demands such as education, health, sanitation, housing construction and so on. In the case of Brazil, for example, between 1976 and 2004, the country saved US$ 60.7 billion by replacing gasoline with ethanol (MACEDO, 2010).

4.9 Ethanol turbine

The vast majority of Brazil's conventional thermoelectric plants run on natural gas, which was consolidated after the Bolivia-Brazil gas pipeline came into operation and rationing at the beginning of the 21st century, making them the first option in the event of a shortage of hydroelectricity.

The *flex-fuel* plant, assuming that ethanol is used most of the time, is an option for Brazil to make its energy matrix cleaner. Due to the country's natural characteristics, hydroelectric plants have the largest installed capacity and are kept in operation most of the time, leaving thermal power plants as an auxiliary method of obtaining energy.

One option for alternative energy to hydroelectricity and keeping the matrix clean is to convert thermal power stations to gas or build new ethanol plants. The north of Rio de Janeiro doesn't have a more viable option than this, as it doesn't have large waterfalls for large hydroelectric plants, if the aim is to maintain the renewable matrix. As will be studied, the region has the conditions to become a bioenergy producer through thermal ethanol plants.

The process of converting or building an ethanol plant consists of modifying the auxiliary systems of a conventional gas-fired plant, few of the turbine's internals are altered. Oxidation conditions and component degradation are also altered, but if there are incentives for research and technology development, this barrier can be overcome.

The turbine converted at UTE-JF is a 43.5 MW LM 6000 manufactured by

General Electric (GE), operating on a simple cycle, but there is the option of turning it into a steam cogeneration plant, significantly increasing efficiency and the total energy generated (MACHADO, 2010).

The internals that needed changing were two injector nozzles and the combustion chamber, which was redesigned to reinforce the one used for NG. The rest of the modifications were for the ethanol supply, such as installing the ethanol supply system, including the pumps, filters, control valves, piping, instrumentation and supervisory system (MACHADO, 2010).

A demineralized water injection system also had to be added to control the level of nitrogen oxide (NOx) emissions into the atmosphere.

Figure 7 shows a cross-section of the LM 6000. You can see the combustor, which is the key part replaced in the conversion. It is the zone where the burning gases are at their maximum temperature.

Figure 7 - Internal view of the LM 6000 turbine and its parts manufactured by General Electric.

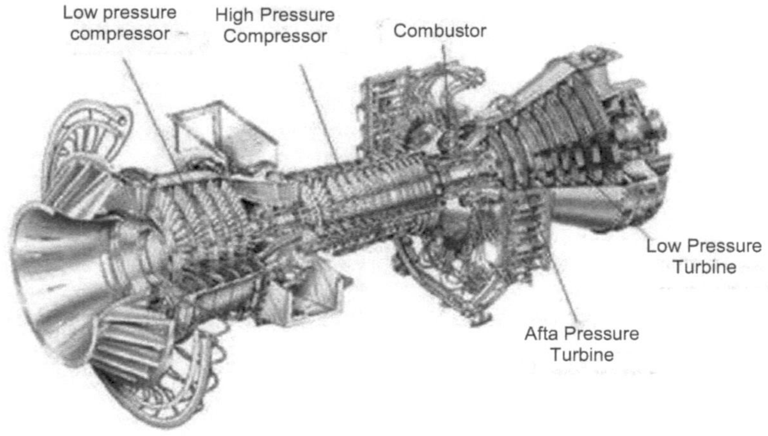

Source: General Electric Power website: http://www.gepower.com

4.10 Implementation of an alcohol pipeline in the NF region

According to Ballou (2006), the movement of products via pipelines is very slow, at no more than three to four miles per hour. On the other hand, it is 24 hours a day, seven days a week, which makes the effective speed much higher when compared to other modes.

The pipeline pumping system is the most reliable, with the lowest operating costs and the lowest risk of loss of production or accidents.

We already have an example of an alcohol pipeline installed in São Paulo, which drains the production of the plants associated with the LOGUM project.

Logum Logistica S.A. is the result of a unique and innovative project by Brazilian entrepreneurs in the engineering, energy and transportation sectors. Six companies - Camargo Corrêa Construgoes e Participagoes (10%), Copersucar (20%), Raizen (20%), Odebrecht Transport Participações (20%), Petrobras (20%) and Uniduto Logistica (10%) - are responsible for bringing together, into a single network, three individual pipeline projects which involved, in addition to pipelines, complex transportation systems by waterway, road, cabotage, as well as operations in waterway terminals, ports and warehouses. (logum.com.br).

A system to interconnect the alcohol-producing plants and transport production more reliably and with less stoppages and setup times, supplying the thermoelectric plant and facilitating the export of the alcohol produced in addition to domestic consumption.

A project carried out by Maciel (2008) designed an ethanol pipeline for the northern region of Rio de Janeiro, where the location of the pipeline was defined using the Center of Gravity Method. Based on Slack *et al.* (1999), the center of gravity method is used to find a location that minimizes transportation costs. The study was based on the 2004-2005 harvests, when there were still 8 mills producing, unlike now when we only have 4 units producing in the NF. We will take some information as parameters for the current work.

4.11 Implementation of the Ethanol Thermoelectric Plant

Brazil's energy matrix is based on hydroelectricity production, with thermoelectric plants being activated when there is low rainfall that hinders electricity production. Conventional thermoelectric plants use natural petroleum gas (NG) which is burned in gas turbine combustion chambers. They are increasingly used due to the country's growing energy demands and the impossibility of building new hydroelectric plants (either for environmental reasons, or because of the generation capacity of the remaining waterfalls that have not yet been exploited), these plants have a high operating cost because the cost of natural gas is high compared to hydroelectric power, which is free (discounting the costs not calculated by the economy, such as environmental damage and the displacement of fauna and native peoples due to the flooding of large areas). A national solution that could benefit not only the country, but which, if we think more conservatively, could help on a global scale, would be to generate electricity by burning hydrated ethanol in gas turbines, instead of NG. A pioneer in the world is the Juiz de Fora thermoelectric plant (UTE-JF), where a turbo generator set was initially converted to ethanol as the primary fuel and NG as the secondary fuel. The Aeroderivative Gas Turbine model LM 6000 from the manufacturer General Electric (GE), with a nominal power of 42 MW, was the object of the transformation. A test of 1,000 hours of operation found that emission levels were lower than those practiced in the plant's operation with natural gas. There was also a reduction in the consumption of demineralized water for operating the turbine with ethanol. The levels of aldehydes emitted by the turbine were much lower compared to the same consumption of an Otto cycle ethanol engine. NOx emissions were reduced by up to 46.7% and there were no SOx emissions.

The improved combined cycle efficiency of the LM 6000-PG and LM 6000-PH can reduce fuel consumption by the equivalent of 33,000 barrels of oil per year, when compared to other similar aeroderivative solutions in their class. GE's LM 6000 also reduces carbon dioxide emissions by 6,500 tons over the course of a year, the same typical operational emissions reduction achieved by removing 2,500 cars from the road annually (ge-energy.com).

Based on these results, it can be seen that by implementing this type of technology on a large scale in Brazil, we will have greater operational flexibility, greater energy security, greater energy diversification and the creation of a new market segment for ethanol in the country and abroad, making it possible to increase sales of the fuel and boosting the business of the sugar-alcohol sector in Brazil.

5. METHODOLOGY

In order to analyze the North Fluminense region's capacity for planting sugar cane and producing ethanol to generate electricity or in the case of surpluses for export, the feasibility of setting up a sugar-energy complex with a thermoelectric plant converted to ethanol was discussed.

The study was carried out through bibliographical research with the aim of obtaining information on the subject and investigating possible obstacles to ethanol production, the implementation of the complex and the export of surpluses. They also discussed issues such as improving the quality of life in the region, including health, income, job creation and replacing fossil fuels with a biomass-based fuel.

The bibliographical research began with the collection of data on the subject. Information on biomass is easy to access and obtain, but data on the machine mentioned, the LM 6000 aero-derivative turbine, is restricted to those who work in the field. In academic circles, only Machado (2010) has technical reports on the application at UTE-JF.

Data on turbine consumption, current ethanol production capacity and tankage capacity in the NF were analyzed, showing the region's maximum potential acquired over time.

With this data, it was possible to carry out the calculations to size the current and related tankage capacity, finally obtaining the table that summarizes them.

Below, in equations 2, 3 and 4, is a description in mathematical language of how the calculations were carried out, followed by a flowchart of the methodological process for obtaining the data in figure 8.

To calculate the consumption of an aeroderivative ethanol turbine, equation 1 below was drawn up:

$$CT = CE \times QH \tag{2}$$

Where:

CT = Turbine consumption;

EC = Ethanol Consumption;

QH = Number of operating hours.

To calculate the equivalent number of turbines, which compares the amount of ethanol produced in Norte Fluminense and the consumption of an aeroderivative ethanol turbine, equation 2 below was drawn up:

$$QTE\ (PE) = P\ (NF)\ /\ CT \tag{3}$$

Where:

QTE (PE) = Quantity of equivalent turbines compared to the volume of ethanol currently produced;

P (NF) = Northern Fluminense production;

CT = Turbine consumption.

In order to calculate the equivalent number of turbines, we can compare this with the tankage capacity, which relates the total possible storage in the NF and how many turbines this volume would be able to supply, in the following format:

$$QTE\ (TC) = CTG\ /\ CT \tag{4}$$

QTE (TC) = Quantity of equivalent turbines compared to the volume that can be stored in tanks in the NF;

CTG = Leverage Capacity in the NF;

CT = Turbine consumption.

The data obtained from these calculations is shown in the table in the Results and Discussions section below.

Figure 8 - Flowchart of the methodological process for obtaining data

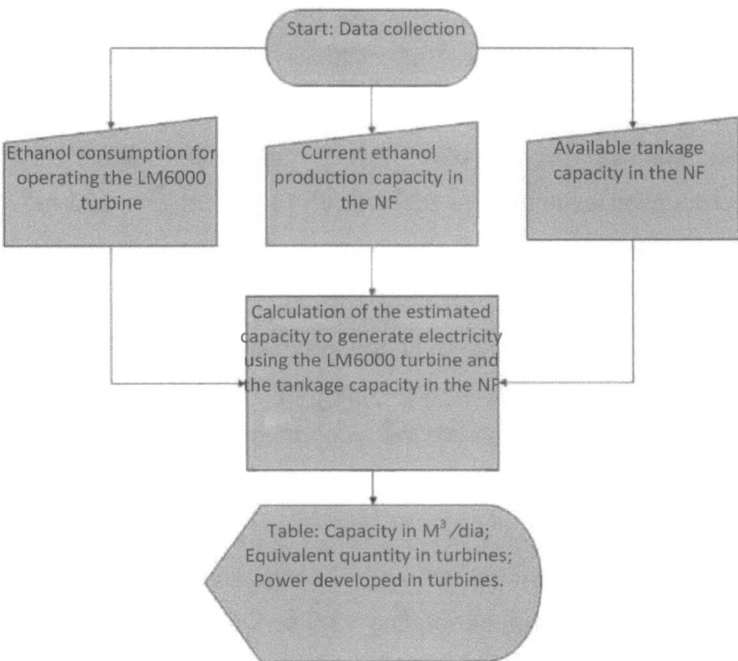

Source: Author

6. RESULTS AND DISCUSSIONS

The bibliographical references section showed the benefits of using ethanol as a primary fuel. We will now show an estimate of electricity production compared to the current reality in the north of Rio de Janeiro.

The Canabrava, Paraiso and COAGRO plants have been authorized to operate, with production estimated at 500 m^3 /d, 130 m^3 /d and 250 m^3 /d, respectively, all of which is only hydrated ethanol, resulting in a production capacity of 880 m^3 /d in the north of the state.

The aforementioned plants belonging to the municipality of Campos dos Goytacazes were analyzed by Maciel (2008), where Paraiso and COAGRO were considered in the project to create the alcohol pipeline. The Canabrava plant was not considered because it was not operating at the time of Maciel's study (2008).

According to Machado (2010), the ethanol consumption for one hour of operation of the LM 6000 turbine, generating 43.5 MW/h is 18400 L/h, so for one day of operation we have 441.6 m3/d for a turbine of this model.

Considering the current production scenario, below is table 10 with data on the production capacity of electricity generated by aero turbines derived from the LM 6000 model.

Table 10 - Estimated electricity generation capacity using the LM6000 turbine currently in the NF.

	M3/d	Equivalent number of turbines	Approximate power developed by the turbines (MWh)
Production in NF	880	1,992753623188406	87
Tank capacity	50.095	113,4397	4915,5

Source: Own authorship.

The table above shows the tankage capacity, data taken from ANP (2016). It shows a reflection of the capacity that the NF region has had over time, i.e. if we consider that the plants that once produced hydrous ethanol were all operating at full capacity, applying the technology in question, we would have a rounded total of 113 turbines operating, generating a total of 4915.5 MWh.

However, it is clear that the amount of ethanol produced in the NF is insignificant compared to other states and that sugar cane planting has been treated as a low priority, giving the false impression that the fuel is not needed in the region. This drop in production that has been happening in the region and in the state is a demonstration that the current governments have no idea of the production capacity and the need to invest in this area in order to generate jobs and income.

Current ethanol production capacity is minimal compared to the logarithmic scale of growth that the region has always shown with sugar and later with alcohol at the beginning of Brazil's energy history.

The region has a large oil basin and large-scale production, but the financial crisis in the country, coupled with the fall in production from older wells and the discovery of the pre-salt basin in various regions along the Brazilian coast, is expanding national production, which is no longer concentrated in this region alone. The likelihood is that there will be an increasing reduction in the value of oil royalties as a result of the decrease in production in the region. The dependence that the region has created on the transfer of royalties is immense and must be overcome as it is withdrawn. The NF has no other concomitant way out of development and the production of cheaper, cleaner energy than the expansion of the bioenergy sector through sugar cane. Sugar cane production must be brought back into the region's matrix, dictated by the pace at which royalty payments fall.

However, the figures presented in the results show that it is impossible for the NF region to operate in this scenario with current production alone. The current figures are not enough to supply the minimum industrial facilities required.

7. CONCLUSION

After analyzing the study, it can be seen that there is no possibility of a project to create an alcohol pipeline network and use it to power turbines for electricity generation unless there is a very high level of investment, In other words, in the country's current economic crisis, private capital must be involved, especially for the installation and maintenance of equipment that is not manufactured in the country. However, this would not make the project any less important or sovereign, since energy is what moves the world and we all need it.

It is a project of the highest value, since it expands the country's energy supply, develops an entire region, generates jobs, income and improves health in the surrounding areas, and also contributes to the reduction of carbon in the atmosphere.

8. BIBLIOGRAPHICAL REFERENCES

AGENCIA NACIONAL DE ENERGIA ELÉTRICA - ANEEL **Average tariff by consumption class and by region**. Brazil: ANEEL, 2016. 1p. Available at: <http://relatorios.aneel.gov.br/_layouts/xlviewer.aspx?id=/RelatoriosSAS/Rel

S

ampRegCC.xlsx&Source=http://relatorios.aneel.gov.br/RelatoriosSAS/Forms/ AllItems.aspx&DefaultItemOpen=1> Accessed on: 29 Feb. 2016

NATIONAL agency for PETROLEUM, NATURAL GAS AND biofuels - anp **ethanol bulletin n 07/2016**. Brazil: ANp, 2016. 27p. Available at: <

http://www.anp.gov.br/wwwanp/images/publicacoes/boletins-anp/Boletim_do_Etanol/Boletim_do_Etanol_No07_JUNHo_2016.pdf> Accessed on: 29 feb. 2016 ANUARIO BRASILEIRO DAS INDÙSTRIAS DE BIOMASSA E ENERGIAS RENoVaVEiS. Minas Gerais: Três, 2015. 128p

BALLOU, R. **Supply Chain Management - Business Logistics**. 5. ed. São Paulo: Bookman, 2006. 616p.

BANCO NACIONAL DE DESENVOLVIMENTO ECONÔNICO E SOCIAL - BNDES **Sugarcane bioethanol: energy for development**. Rio de Janeiro: Organized by BNDES and CGEE, 2008. 316p.

BIOCOMBUSTIVEL, C. P.; SMES, **Food security associated with the production of biofuels**. Especial Pàginas Verdes, Brazil, v.1, n.4, p.1-4, 2014. Available at:

<http://portalpetrobras.petrobras.com.br/conteudo/petr_banco_anexos/a_pbio /paginas_verdes_edicao_4_27_06_14.pdf> Accessed: Feb. 19, 2016

BIOSEV - Sugar and **Alcohol Sector**. Available at: <http://www.mzweb.com.br/biosev/web/conteudo_pt.asp?idioma=0&conta=2 8&tipo=30884> Accessed on: Feb. 29, 2016

CENTER FOR ADVANCED STUDIES IN APPLIED ECONOMICS - CEPEA/ESALQ, **Weekly Hydrated Ethanol Indicator** . São Paulo: CEPEA,

2016. 1p. Available at: <http://cepea.esalq.usp.br/etanol/> Accessed on: 29 Feb. 2016

NATIONAL SUPPLY COMPANY - CONAB. **The Foundations of the Crisis in the Sugar and Alcohol Sector in Brazil.** Available at: <http://www.conab.gov.br/OlalaCMS/uploads/arquivos/abb7ef86c87e6c8486 32fed87188cdb7..pdf> Accessed on: March 3, 2016.

DEPARTMENT OF SUGARCANE AND AGROENERGY **Sugar and Alcohol in Brazil.** Brasilia, DF: SPAE/MAPA, 2006. 250p.

ENERGÈTICA, E. P.; ENERGIA, M. M. E. **Thermoelectric Generation - Biomass.** National Energy Plan 2030, Brasilia, n.12, p.49, 2007. Available at: <http://www.epe.gov.br/PNE/20080512_8.pdf> Accessed on: Feb. 19, 2016

FISCHETTI, D.; SILVA, o. **Etanol: a revoluçâo verde e amarela.** 1st ed. Săo Paulo: Bizz Comunicaçăo e Produçôes, 2008. 264p.

FUSER, I. **Energy and International Relations.** 2nd ed. Săo Paulo: Saraiva, 2013. 240p.

LOGUM, 2016. Available at: <http://www.logum.com.br/php/o-sistema-logum.php>. Accessed on: 29 Feb. 2016.

GENERAL ELECTRIC COMPANY - , G. E. C. - **Fast, Flexible Power: Aeroderivative Product and Service Solutions.** GE, Houston, Texas, USA, n.1, p.1-16, 2013. Available at:

<https://powergen.gepower.com/content/dam/gepower-pgdp/global/en_US/documents/product/aeroderivative-products-services-brochure.pdf> Accessed: 25 Feb. 2016

KLOSS, E. C. **Transformation of ethanol into a commodity: Perspectives for Brazilian diplomatic action.** Brasilia: FUNAG, 2012. 232p.

MACEDO, I. C.; SOUZA, E. L. L. **Ethanol and bioelectricity: sugarcane in the future of the energy matrix.** 1. ed. Săo Paulo: Luc Projetos de

Comunicaçâo, 2010. 314p.

SILVA, Roberto Cezar Rosendo Saraiva da; CARVALHO, Ailton Mota de. Economic formation of the Northern Fluminense Region. In: PESSANHA, Roberto Moraes; NETO, Romeu e Silva (Orgs.). Economy and development in the North of Rio de Janeiro: from sugar cane to oil royalties. Campos dos Goytacazes, RJ: WTC, 2004. p. 27-72

SUGAR CANE'S: Twelve studies on Brazilian sugar cane. São Paulo: UNICA, 2007.   230p.   Available   at:

<http://www.google.com.br/url?sa=t&rct=j&q=&esrc=s&source=web&cd=2&c ad=rja&uact=8&ved=0ahUKEwjlyL_ViLzMAhUFH5AKHZGECSkQFgghMAE& url=http%3A%2F%2Fsugarcane.org%2Fresource- library%2Fbooks%2FSugar%2520Canes%2520Energy%2520- %2520Full%2520book.pdf&usg=AFQjCNFCb63tXyn-Cv2c- wGlar7hOlwLcw&sig2=huHoqTCqreJkXZDe8qr3jg> Accessed: May 2, 2016

MACHADO, Joăo Mendes. **Conversion studies of an aeroderivative natural gas turbine for dual-fuel operation:** natural gas and ethanol. 2010. 75f. Monograph (Specialization in Technical Plant Supervision) - University of São Paulo, São Paulo - SP, 2010.

MACIEL DA SILVA, Gustavo. **Study of the implementation of a pipeline system for the export of ethanol produced in Campos dos Goytacazes**. 2008. 57f. Monograph (Bachelor's Degree in Production Engineering) - Universidade Estadual do Norte Fluminense Darcy Ribeiro , Campos Dos Goytacazes - RJ, 2008.

MINISTÉRIO DA AGRICULTURA, PECUARIA E ABASTECIMENTO **Balanço Nacional da Cana-de-açùcar e Agroenergia**. Brasilia: MAPA/SPAE, 2007. 139p. Available at:

<http://www.agricultura.gov.br/arq_editor/file/Desenvolvimento_Sustentavel/A groenergia/estatisticas/PDF%20-%20BALANO%20NACIONAL_0_0.pdf>

Accessed: Feb. 19, 2016

MINISTRY OF MINES AND ENERGY **National Energy Balance 2014 - Synthesis Report**. Rio de Janeiro: EPE - Empresa de Pesquisa Energètica, 2. 54p.

NETO, R. E. S.; PESSANHA, R. M. **Economia e desenvolvimento no norte fluminense: da cana-de-açùcar aos royalties do petróleo**. Campos Dos Goytacazes: WTC Editora, 2004. 364p.

NOVACANA. **Technological development in the sugar-energy sector**.

Available at: https://www.novacana.com/estudos/o-desenvolvimento-tecnologico-do-setor-sucroenergetico-241013/. Accessed on: 15 Feb. 2016.

NOVACANA. **Benefits of using ethanol**. Available at:

< https://www.novacana.com/etanol/beneficios/>. Accessed on: 15 Feb. 2016.

SOARES, A. P. **O Sector Sucroalcooleiro e o Dominio Tecnològico** São Paulo: Naippe/usp, 2015, 29p.

VASCONCELOS, G.F. **O poder dos trópicos: Meditation on energy alienation in Brazilian culture**. 1. ed. São Paulo: Sol e Chuva, 1998. 304p.

VIDAL, J. W. B. Brasil: Civilização suicida. Brasilia: Star Print Gràfica e Editora, 2000. 88p.

. The splintering of the nation. 1. ed. Petrópolis: Vozes, 1995. 216p

SUGAR CANE INDUSTRY UNION - ÙNICA. **Biomass can guarantee 13% of the country's energy generation**. Available at: < http://www.unica.com.br/na-midia/27455300920333453814/biomassa-pode-garantir-13-por-cento-da-geracao-de-energia-do-pais/> Accessed on: Feb. 19, 2016